D1766455

COLEG CAMBRIA

Llysfasi

Learning Zone – Parth Dysgu

01978 267917

Accession No. 986 R ------------------- **Rhif Derbyn**

The Combine Harvester

11/20

INE

ER

First published 2018

Amberley Publishing
The Hill, Stroud,
Gloucestershire, GL5 4EP

www.amberley-books.com

ISBN: 978 1 4456 7771 2 (print)
ISBN: 978 1 4456 7772 9 (ebook)

British Library Cataloguing in Publication Data.
A catalogue record for this book is available from the British Library.

Typeset in 10pt on 13pt Celeste.
Typesetting by Amberley Publishing.
Printed in the UK.

Contents

Introduction

The combine harvester is perhaps one of the most important machines to have ever been devised by man. That is quite a statement but one that is justified by the simple fact that, without the combine harvester, the world would struggle to be able to feed itself. The global wheat crop is so important that the rise and fall of prices on the world market affect everybody on the planet.

The beauty of the combine harvester is in the fact that it has combined what were once two completely different tasks – those of harvesting and then threshing the grain, which it does all in one operation, out in the fields. This is indeed what gives it the very name of combine harvester!

This book is the story of the combine harvester, how it developed from the two separate binder and threshing drum machines and how it then developed into a machine that was used worldwide for a whole host of crops. It is a big story and in a volume of this size much has to be missed out; however, as we shall see, the basic principles of these machines have not changed a great deal over the decades. Rather, what has changed is the level of sophistication, with less and less input being required from the operator. There are also three different threshing mechanisms used in modern machines, which use either the tried and tested straw walkers, an intrinsic part of the original threshing drums, or the rotary type developed during the 1970s. More recently hybrid types have appeared, offering the best of both the earlier methods.

What has not changed is just how important the combine harvester still is to everybody in the world today. They have also grown in size, particularly in recent years, and the largest combine harvesters are now among the biggest machines ever built.

In the following story we also find out that much of the development of the combine has been down to just a few manufacturers, many of who, despite takeovers and mergers, still produce these machines to this very day. And, although the combine harvester was born in the United States of America, today it is truly a global phenomenon.

The following is this fantastic machine's story from the point of view of its use in the British Isles: a story that begins with horses and finishes with self-driving mega machines. There is probably no other machine that has made such a difference to the world at large as the combine harvester – it is indeed a modern miracle!

Wheat, one of the most important crops grown by farmers and a driving force in the evolution of the combine harvester.

The first combine harvesters were large and pulled by horses. Later, smaller versions were developed that could be pulled by small tractors, such as this Allis-Chalmers All Crop 60, which is being pulled by a D272 tractor of the same make. (Photograph: Kim Parks)

Massey-Harris proved the concept of a self-propelled combine with the arrival of the Model 21. (Photograph: John Chapman)

Two Claas SF combines waiting for their next harvest.

British farmers adopted self-propelled combine harvesters by the 1950s and 1960s, with British makes such as Ransomes rubbing shoulders with the likes of Massey Ferguson. (Photograph: Kim Parks)

The USA was the spiritual home of the combine harvester and machines such as this silver Gleaner were very popular during the 1970s and '80s. (Photograph: Paul Reeve)

A line of Canadian-built Massey Ferguson 865 combines.

Claas Lexion combines were sold in North America for a time in Caterpillar colours and were built in the USA. Note the rubber track units replacing the front wheels. (Photograph: Paul Reeve)

Vintage working days give the public a chance to see combines from all eras at work.

Combines today are big – take this Claas Lexion 600 Terra-Trac with a 40-foot-wide header harvesting oilseed rape, for instance.

CHAPTER 1

A Tale of Two Machines

The cereal harvest has always been seen as one of the most important times of the year and is therefore crucial to the success of any farm, and of those that work on it. It was also a very labour intensive task to gather in the harvest when everything was done by hand. Teams of men would use scythes to cut the straw of the wheat, or whatever cereal crop was being harvested. The crop would then be picked up by hand and made into small piles, or 'stooks', in the field, where it could dry out for a couple of weeks before being collected, once again by hand, and then made into a weatherproof stack where it would be stored until the winter months.

The process was only half complete at this stage and in the winter the stacks would be dismantled and the long job of threshing the grain from the heads would begin. This was done once more by hand power, as flails were used to bash the crop on a dirt floor of a barn, preferably one with a good air flow passing through to assist with the separation of the grain from the husks and other waste material. Only then could the grain be scooped up and stored, ready for use or to be sold.

Mechanisation eventually began to arrive, with the sail reaper being the first major step forward, as it was a horse or oxen-powered device that not only cut the crop, but also left it in groups to make 'stooking' easier. This machine eventually evolved into the binder, which could actually produce tied sheaves of corn, leaving these behind it on the ground as it worked across the field. The self-tying binder was a real revolution and enabled the crop to be cut much more quickly than had ever been possible before.

Gradually the horse gave way to the tractor as these new machines became ever more popular during the twentieth century. This was hastened by the fact that the binder was a heavy tool that required a minimum of two horses, and preferably three, to haul it across the field when in work. Additionally, as all its mechanisms were driven by a land wheel, it was tough work during the heat of summer and the binder soon got the rather sinister nickname of the 'horse killer'!

The tractor-drawn binders worked in exactly the same way as the horse-drawn machine, and indeed many farmers adapted their horse-drawn implements to be pulled by a tractor. In work, a reciprocating knife cut the crop and revolving sails helped make sure it was placed onto the canvas conveyor, which took it up to the platform above the rear wheels where the crop was bunched, a knotter then tying them before the completed sheave was ejected onto the cleared ground. It sounds simple but watching a binder at work is still a marvellous thing to this day, and when people saw them for the very first time it must have seemed nothing short of magic.

With every country often having their own native makes and models, from the early days McCormick and Deering were pioneers in the United States, along with Massey-Harris in Canada, while in Britain the Albion and Woods binders were very popular, among several others. McCormick and Deering were competitors in the building of farm implements, including reapers and later binders, with Cyrus McCormick introducing his first reaper in 1831. Then, in 1902, the two companies merged together, along with three other firms, to form the International Harvester Company.

Massey-Harris also had its origins in two businesses, with one being founded in 1847 by Daniel Massey and the other being the Harris company, with which it merged in 1891. Daniel Massey had been responsible for making some of the very first mechanically operated threshers and reapers, and the new Massey-Harris firm was based in Toronto, Canada.

Cutting the corn was still only half the job, however. On the threshing side, special threshing drums were gradually developed that could be fed with sheaves of corn at one end, with a large drum then threshing out the grain from the ears and separating the chaff. The grain would emerge at the far end to be put into sacks while straw walkers moved the straw around on its way out of the threshing drum to collect any extra grain that had not been removed previously, which was then returned to the sack filling area. Threshing drums were one of the first things to be turned over to steam power, portable steam engines being used to provide the power necessary via a belt drive. Self-propelled traction engines gradually replaced the portable engines, followed by tractors taking over as the twentieth century progressed.

North American threshing drums built by the likes of International Harvester were of the peg drum type, which was eminently suitable for the tough wheat grown on the prairies, but a slightly different design was used in countries such as Britain, with more use being made of straw walkers due to the longer and damper straw itself. To name but a few, firms such as Clayton & Shuttleworth, Marshall, Ransomes and Garrett were familiar names on threshing drums in Britain during this period.

These large threshing drums must have been as wondrous a sight to the country folk as the binders had been. Being able to put the complete crop into one end and have it separated into its constituent parts mechanically must have seemed very impressive. The chaff could be collected in sacks for cattle feed, the grain was sacked up ready to go to the merchants or into store, and the straw came out the rear to be made into a stack for winter livestock bedding; alternatively, it could be made into large bales by a stationary baler, which was also powered by the traction engine. All in all, it was a very efficient system.

The pattern was soon set for cutting the cereal crops in late summer with a binder and then making a stack that would be threshed by a contractor's threshing drum and tackle, which visited all the local farms during the winter. This would be the way it stayed for many years, particularly in Britain, with farms still using the binder and threshing drum up to the 1960s. But change was in the air and even by the early years of the twentieth century a combined harvester and thresher was beginning to appear on the wide-open prairie lands of North America.

The first mechanical cereal harvester was the sail reaper. Originally horse-drawn, many were later adapted to be pulled by early tractors.

The reaper was replaced by the binder, which not only cut the crop, but also made neat bundles called sheaves. It was originally pulled by real horse power.

Tractors gradually replaced the horse as the motive power for the heavy binder. .

This binder was built by Massey-Harris and the canvas that transports the cut corn to the knotter can be clearly seen.

After the sheaves have been tied, they are thrown from the side of the Massey-Harris binder.

Sometimes a hay sweep, like this front-mounted example, would be used to collect the sheaves of corn and take them to the stack.

Most often the stooks of sheaves would be picked up by hand a couple of weeks after cutting, before taking the crop to the stackyard.

Threshing would usually take place in the winter months and contractors did the rounds of the farms in turn. This threshing drum was built by Foster, based in Lincoln.

North American threshing drums looked distinctly different to their British counterparts, as shown by this International Harvester example. (Photograph: Kim Parks)

A Ransomes threshing drum, built in Ipswich, showing the large number of belts used to drive it, usually by a steam-powered traction engine.

Tractors replaced steam power and the Field Marshall, with its single cylinder diesel engine, was particularly good at this task.

CHAPTER 2

Trailed Combines

Even the early reaper-threshers, as they were called (the term combine harvester not yet being coined), were huge machines. Built by the likes of Holt and Case, sometimes as many as a hundred horses were needed to pull these monster machines through the wheat crops.

Made out of steel and wood, these leviathans were in many ways pretty advanced, taking the latest threshing technology used in stationary threshing drums and attaching a wide cutting header complete with wooden sails to gather in the crop as it was drawn across the land. The grain would end up in a large holding tank and the straw and chaff would be directed out of the very rear of the machine.

As the twentieth century progressed, tractors instead of horses would pull these mighty machines, but they still relied on their own on-board engines to provide the power needed to drive all the working components.

It would be these reaper-threshers that would evolve into combine harvesters, especially when more and more manufacturers began producing their own machines, such as International Harvester and Massey-Harris. Gradually they also began to produce smaller versions, even to the point where the tractor's own power take-off could provide the power to run all the threshing and cutting mechanisms.

Some of the large combines were shipped to Britain in the 1930s and after, but they only enjoyed fairly limited success as they were not designed for the damper conditions and longer straw found on this side of the Atlantic. Plus, not many farms were large enough to justify such huge machines.

The smaller trailed combines were a different story, however, and began to become quite popular. Once again American makes were the main contenders, with the likes of John Deere, International Harvester, J. I. Case and Allis-Chalmers all competing with each other for sales. The early crude designs were soon replaced with much lighter and more modern looking and efficient combine harvesters.

Allis-Chalmers enjoyed great success with the All-Crop trailed combine, which was powered by a version of its own Model B tractor engine. This was a well-developed machine that, as its name suggests, was ideal for harvesting a range of different crops as well as cereals. Eventually these would be built in the UK as well as North America.

International Harvester produced a range of smaller trailed combines that were very popular, versions of which were eventually built in the Doncaster factory in Yorkshire after the Second World War. Other combines were also produced in Germany and France.

In Europe, German firm Claas, begun in 1913 by the five Claas brothers, had been working on combine harvester development after much success with binders. The first was a machine that was constructed around a German-built Lanz Bulldog tractor, which was produced from 1930, before being replaced by the more conventionally designed trailed MDB in 1936. The larger Super design was in a prototype stage during the Second World War but entered full production afterwards. These distinctive machines were painted all silver and the early ones actually used the large tyres from bomber aircraft to support them. Unlike the American designs, the Claas Super came with an easily folding header, making it much easier to transport along narrow roads from field to field. It was also developed with European crops and field conditions in mind and so it was arguably able to produce a much better sample than the American machines.

British firms even jumped on the bandwagon, with Clayton producing copies of the large American machines, most of which were exported rather than sold on the domestic market. The long-running tractor and threshing machine builder Marshalls of Gainsborough produced the trailed Grain Marshall combine harvester. Ransomes, Sims & Jefferies, based in Ipswich, also produced trailed combines built under license from a Swedish firm, beginning with the Bolinder-Munktell MST42 in 1953, followed by the MST56 in 1959.

Canadian Massey-Harris was another name that would become well known for its combine harvesters. This business had been a pioneer in binder design and was also a large builder of reaper-threshers before moving into combine harvester design, but it would be one particular innovation that would see the Massey-Harris name at the forefront of combine harvester design.

In the USA, Holt was one of the pioneers of early combine harvesters such as this massive Model 36. (Photograph: Gerry Hutchinson)

The Holt Model 36 was a heavy machine and not ideal for wetter European conditions. On the plains of the Midwest, however, it was a very different story. (Photograph: Gerry Hutchinson)

International Harvester was also an early American builder of combine harvesters, or reaper-threshers as they were widely known as at the time. Note the wide side-mounted header with the canvas belt to take the crop up to the threshing mechanism. (Photograph: Gerry Hutchinson)

Smaller combine harvesters brought the technology to smaller farms. International Harvester made many smaller machines that were ideal for pulling behind the tractors of the time.

The smaller size of the new breed of combine harvesters made them much more manoeuvrable, and thus ideal for smaller size fields. A tank was also often fitted for the bulk collection of the grain.

John Deere produced smaller combines in the USA, such as this Model 12A from 1940, some of which found their way to Britain.

The John Deere combines had a very curved design, something of a trend in small trailed combines by several manufacturers. This one is fitted with a bagging station, as was the norm for machines destined for Britain.

Allis-Chalmers had a great deal of success with its All-Crop combine in the firm's native USA and also in Britain. The first All-Crop arrived in 1935, followed by the 60 in 1940.

The Allis-Chalmers All-Crop 60 was powered by its own petrol engine. It had a tall and sloping canvas belt to lift the crop up into the cross-mounted threshing drum at the rear of the machine.

Another All-Crop 60 being pulled by an Allis-Chalmers Model B with a Perkins diesel engine.

The Claas Super was an impressive looking machine. Built in Germany, it was developed during the Second World War and launched in 1946.

The British grain handling system was based around sacks for many decades and so most farmers opted for bagger versions of combines like the Claas Super.

The large tyres fitted to the Claas Super combines were originally sourced from Second World War bomber aircraft.

The Claas Super was a very neat design, with a folding header for road transport and a very compact threshing drum. At the rear a buncher was fitted to tie the straw into neat bundles to make later collection easier.

The Aktiv combine from Sweden was a very successful design in its native Scandinavia and several were imported into the UK.

Ransomes built versions of the Swedish Aktiv combines under license at Ipswich.

CHAPTER 3

Going Self-Propelled

In fact, it was Massey-Harris that was responsible for changing the design of the combine harvester forever! One of the biggest problems with pulling a combine harvester behind a tractor was that there had to be enough room to allow the tractor to run on harvested ground and the tractor also had to be offset to the combine reel. This was fine once the field had been opened up, but meant that the first run around the field's headland had to be done by hand and cleared away to allow room for the tractor to pull the combine around for the first time.

The obvious way around this problem was to make the combine self-propelled and keep the header out front, the rest of the machine running behind on harvested ground. This also had the advantage of putting the operator in a much better position to watch the cutting operation, as the header was now laid out at the very front of the machine.

The Massey-Harris 20 was the result; the world's first self-propelled combine harvester, it first saw the light of day in 1938. As this was a rather tall and top heavy machine, it was redesigned as the self-propelled Massey-Harris 21 in 1940, and it then changed everything forever.

Powered by a six-cylinder Chrysler engine, the front-mounted header cut the crop and placed it on a canvas table before feeding it into the threshing drum via a central conveyor. After being separated the grain was stored in a large tank, which had its own unloading auger. Later, the 21A used an auger instead of the canvas table and also made use of an electric mechanism for raising and lowering the header table itself.

As part of an attempt to prove that the self-propelled 21 was a viable proposition, and also to help with the need to harvest as much food as possible during the war, Massey-Harris launched the Harvest Brigade in 1944 using a large team of 21 combines to harvest across the country, from south to north, following the ripening grain. This was a huge success and no doubt was responsible not only for the large-scale acceptance of the self-propelled combine by American farmers and contractors, but also the success of the Massey-Harris company itself.

Other firms were close behind, with Case following very quickly with their own self-propelled model and all the others soon following suit, most notably International Harvester and John Deere. The American farmers took to the idea and soon a great deal of the nation's cereal crops were harvested by self-propelled combine. However, some still preferred trailed combines and most manufactures continued to produce trailed versions.

It was the Massey-Harris 21 that was the British farmers' first glimpse of a self-propelled machine when several were imported during the Second World War, delivered as a kit of parts in large wooden crates. These machines did very well considering they were not designed for British conditions and crops, but many farmers struggled with the bulk grain tanks. In Britain the handling of grain was very much tied up with sacks and all the grain was moved in this way around the farm, on trucks on the road and even on the railways.

The answer was to fit bagging platforms to the combine itself, with another man being able to remove full bags and place empty ones onto the machine as the crop was harvested, rather than the grain going into a tank. Bagger models of combine harvesters remained popular in Britain well into the 1960s.

Claas was soon producing a self-propelled combine from 1952 – the SF, which was originally called the Hercules but had to have its name changed as another company had already registered it. The SF proved to be a very good machine and started this German firm's entry into self-propelled combines in such a way that it would eventually become a market leader. The SF was a large machine at the time with three choices of cutter bar width, from 2.4 to 3.6 metres, and with either six-cylinder 60 hp or 56 hp four-cylinder petrol engines, plus the option of either a bagging platform or a bulk tank. Such was its success that it remained in production right up to 1963.

Claeys was a small firm from Belgium but it also enjoyed a great deal of success with the self-propelled combine harvester. Leon Claeys built his first threshing machine in 1906 in a factory at Zedelgem and by the 1950s this factory was producing trailed combines. The first Claeys self-propelled, the MZ, arrived in 1952. Claeys was taken over by the New Holland company from Pennsylvania, USA, in 1964 and combine harvesters were then built in the USA as well as in Belgium for both markets and for the rest of the world.

Ransomes also entered the self-propelled market in Britain, producing a range of machines in its blue livery, starting with the 902 in 1956 and built at the firm's factory in Ipswich, Suffolk. These increased in size as time passed and also included several innovations, such as a pre-threshing cylinder by the time the Cavalier arrived in 1966, although combine production came to an end in 1976.

Marshall, however, had very little success with its self-propelled combine in 1954, and it was quickly dropped from the price lists.

John Deere, who had sold several trailed combines in Britain during the war, produced its first self-propelled combine, the 55, in 1946, but had lost its presence this side of the Atlantic. This was remedied after buying the German firm Lanz, based in Mannheim, and in 1956 both tractors and combine harvesters were once again offered in Europe and the UK. Self-propelled combines were included in this line-up with the 330, 430 and 530 models.

Allis-Chalmers Gleaner combines were built at Essendine in Rutland, but failed to really take off long term, unlike the earlier All-Crop trailed machines, while International Harvester combines built in Germany were sold into the UK marketplace with some success.

Claas built on the success of its SF combine and introduced the Matador in 1961, a large machine that eventually spawned two versions, the Standard and the Giant. Based on the earlier machine, but beefed up and with new features, the Matador was a stunningly successful machine, which was originally still painted in the distinctive silver paint used by this company, until green took over later in its production life. Several cutting widths were available and most were fitted with Perkins diesel engines.

In Sweden the Bolinder-Munktell firm produced its first self-propelled combine, the MST91, in 1950. Volvo took over the firm and continued producing combine harvesters, including those built by Aktiv. Volvo eventually left the agricultural market in 1980 but the Aktiv firm continued to build combine harvesters.

Meanwhile, Massey-Harris started building combine harvesters in Kilmarnock, Scotland. After the success of the 21, various newer American models were adapted to British conditions and were produced in this factory, starting with the 722, which was built in Manchester, and then the 726 in Scotland. Most were of the bagger type, although a bulk tank was also offered as an option. Engine choices included either a TVO-fuelled six-cylinder Austin Newage power plant or a Morris petrol engine.

The 780, introduced to the UK in 1954, was a truly successful machine and bridged the gap of the merger between Massey-Harris and Ferguson, being branded as a Massey Ferguson 780 from 1958, by which time the 780 Special featured an improved specification. Austin Newage TVO or Perkins L4 diesel engines were the main power options and many farmers were still specifying bagging off platforms.

The Massey-Harris 735 of 1956, soon rebadged as a Massey Ferguson, was a small combine designed for use on farms that could not otherwise justify the larger machines. Most were again of bagger type but a grain tank was offered and, although not capable of high outputs, these little combines were very popular in bringing cereal harvest mechanisation to even the smallest of farms.

The 780 was eventually replaced by the 788 with more features, but the real major development was the launch of the 400 and 500 combine harvesters from Kilmarnock in 1962 with an all-new saddle-tank design. Similar combines were also produced in Canada for sale throughout North America and these proved very popular with the large contractors, or custom cutters as they were known, who followed the ripening corn across the Midwest.

From the mid-1960s the improved 410, 415, 510 and 515 were perhaps the most popular combines in Britain and included such new features as the Multi-Flow re-threshing system on the models ending with a number 5. Power on the biggest was now 104 hp.

New Holland combines also bore the Clayson name in Europe from 1969 and continued designs first started by Claeys. Soon much larger machines were being built as the brand continued to grow, and combines would continue to be built in Belgium, and later in the United States, still in the distinctive Claeys yellow livery. By the 1980s the New Holland name would take over completely and the legacy of the Claeys business was lost, except for the yellow paint. From 1971 new models were introduced such as the 1530, which was the replacement for the earlier M133 model. The new machines kept the central driving position and proved very popular, including in Britain.

The self-propelled combine did not have it all its own way though, as most manufacturers were still offering a limited range of trailed types right into the 1970s. The JF company from Denmark came up with the novel idea of building its combine around a conventional tractor, allowing the tractor to be taken out and used for other tasks after harvest. First sold in the early 1960s, different models were produced through the 1970s.

One method of making a self-propelled combine was to make a wraparound machine, in this case a Gleaner built around a Fordson F tractor. The problem with this, and the later Ferguson attempt, was that the driver had very poor vision, hidden away as they were within the combine!

The Massey-Harris Model 21 was the first commercially successful combine harvester and set the trend for future machines. (Photograph: John Chapman)

A rear view of two Massey-Harris 21 combines at work, clearly showing the round grain tanks. (Photograph: John Chapman)

The operator had a superb view of the full width of the cutter bar on the Massey-Harris 21, as well as being able to see when the grain tank was full by simply looking behind him. (Photograph: John Chapman)

Combine production in Canada evolved from the 21 into machines such as this Super 27, which is seen cutting wheat in Norfolk after being imported for a private collection.

From the back, the Massey-Harris Super 27 is a distinctive looking machine, and the unusual rotary straw spreader is a major feature at the very rear of the combine.

The Claas SF was in many ways the European answer to the concept of a self-propelled combine and was originally launched as the Hercules in 1953. (Photograph: Kim Parks)

The SF was available with a choice of 60 hp six-cylinder diesel or four-cylinder 56 hp petrol engines. (Photograph: Kim Parks)

This is an example of the later Claas SF B, showing the mass of pulleys and belts that were essential for its operation.

The driving platform of the Claas SF B, showing the basic seat and the main dials mounted to the right on the handrail.

Above: Compared with the Massey-Harris 21, the Claas SF B stood very high.

Left: Taken back in the 1970s, this picture shows a Claeys M103 imported into Britain by Bamford at work in a crop of barley in East Sussex. (Photograph: Kim Parks)

Right: First introduced in 1958, the M103 was powered by a Ford 80 hp diesel engine mounted up high behind the grain tank. (Photograph: Kim Parks)

Below: A preserved Claeys M103 at a show in Ireland – just one of these very popular machines that still survive.

Above: A later
Clayson 122
showing the newer
square styling
introduced on new
models after Claeys
was taken over by
New Holland.

Left: Seen at work
in the late 1970s,
this New Holland
Clayson M133 is
equipped with a
cab to give the
operator some
protection from
the dust and dirt.
This was somewhat
negated by the door
being kept wide
open to let the heat
out! (Photograph:
Kim Parks)

Ransomes replaced their first self-propelled, the 901, with the 801 in 1963. (Photograph: Kim Parks)

The Ransomes 801 was powered by a Perkins 42.5 hp diesel engine mounted up behind the grain tank. (Photograph: Kim Parks)

The Ransomes 1001 combines built from 1964 came with the choice of either 62 hp four-cylinder, or 90 hp six-cylinder, Ford diesel engines. (Photograph: Kim Parks)

Claas launched the Matador in 1961. Dwarfing the SF, the Matador was extremely popular throughout Europe.

A rear view of the Matador Giant, as it later became known, shows just how big this machine really was, especially in the 1960s.

Silver grey paintwork was finally replaced by a new green colour on the Matador from 1963.

The driving station of the Matador closely followed that of the SF, although the option of power steering was much appreciated by operators.

The engine, an 87 hp Perkins six-cylinder, was once again perched high at the back of the grain tank.

A smaller version of the Matador called the Standard was offered from 1962 to meet the requirements of those who did not need the huge output of the Giant version.

Claas also built a range of smaller combines to cater for all sizes of farms and these included the Columbus from 1959 with its side-mounted engine. It was usually fitted with a bagging off platform.

The Claas Mercury was introduced in 1963 to fill the gap between the Columbus and the firm's larger models and was powered by a Perkins 52 hp four-cylinder engine. (Photograph: Kim Parks)

From Sweden, Aktiv entered the self-propelled market producing a combine based on similar components to those used in its trailed versions.

Volvo had a long history of making threshing machines and, with the creation of Volvo BM and the acquisition of the makers of the Aktiv combines, these machines became even more popular in other European countries, including the UK.

The Massey-Harris 722 was the first British-built combine from this manufacturer and was built at Manchester.

With its TVO-fuelled engine mounted down low within the workings of the machine, the 722 had a low centre of gravity, making it more stable on slopes.

Massey-Harris became Massey Ferguson in the 1950s and the 780 model, built in Kilmarnock, bridged the transition. (Photograph: Kim Parks)

The improved 780 Special was available with either an Austin Newage petrol/TVO or a Perkins L4 diesel engine.

Even if a farmer chose a tanker version of the 780, this method would be used to transfer the grain into sacks for storage and distribution.

So popular was the Massey Ferguson 780 model in Britain that after production ceased in 1962 the 788 replaced it in 1963.

Basically a restyled 780, the 788 featured a horizontal air-cooling intake and a new straw walker hood, as well as the addition of safety guards.

The Massey Ferguson 735 was introduced for smaller size farms, was powered by a four-cylinder petrol engine, had a 6-foot-wide cut and could be had with a small grain tank or as a bagger model.

Right: The Massey
Ferguson 400 and 500
models introduced
in 1962 proved to
be very popular.
For the British
market they were
built in Kilmarnock.
This is a later 415
model at work in
East Sussex in the
1970s. (Photograph:
Kim Parks)

Below: Seen emptying
on the headland of a
Suffolk field in 1985,
the Massey Ferguson
515, launched
in 1966, was the
flagship model.

A Perkins 104 hp diesel engine provided the power to run the 515.

The Multi-Flow system was mounted on the bottom of the rear hood and consisted of a rotary beater to re-thresh the straw and an elevator to take any grain found back up to the grain tank.

The New Holland Clayson 1530 was launched in 1971, with power coming from a 113 hp Ford engine. It could be fitted with headers up to 13 feet wide.

The dust flies as a New Holland Clayson 1530 cuts a crop of oats in East Sussex.

CHAPTER 4

Mod Cons and Giants

Claas brought out the Senator range in 1966 to replace the existing Matador models and soon a range of these machines, including the Protector and Mercator, were produced. All featured completely new styling, with square side panels enclosing the mechanisms and providing a more modern, if rather boxy, profile.

Gradually cabs became an option on the Senator models, although these were fitted at extra cost by the dealer before reaching the purchaser and were built by outside firms. Ever since the first self-propelled combines had appeared, the operator sat out in the open, exposed to all the dust and dirt associated with the cereal harvest. The addition of a cab improved the driver's lot immensely, but these metal and glass boxes soon heated up during the hot summer months when they were most often used, and so air-conditioning quickly became a necessary addition.

All the various manufacturers also offered cabs, but these were, as with the Claas machines, very much add-on items at the point of sale. Gradually, this would change and cabs would become an intrinsic part of the combine itself.

The Senator machines had seen engines of 105 hp fitted and cutting widths of up to 12 feet, but the Dominator range that replaced them from 1970 would see bigger engines and wider cutting widths. By the time the 6 Series was launched in 1978, a brand-new purpose-built cab was a main feature of the new range, and the largest, the 106, was powered by a Mercedes-Benz diesel engine while also benefitting from five straw walkers.

An unusual combine of the 1970s was the Lely Victory. This was a large combine for its day with headers up to 18 feet wide, this being made possible by the clever idea of the header folding up hydraulically into two vertical sections for transport. This meant that it could fit down narrow country roads and through field gateways with ease, and the press of a button would lower the two header sections in an instant, ready for work.

Built under license by Fisher Humphries, the threshing machine manufacturer from Wiltshire, the design of the Victory belonged to the German firm Dechentreiter. Lely bought Fisher Humphries in the middle of the 1960s – hence the Victory combine gaining the Lely name as it was produced from 1965 to 1980, with both Mark II and Mark III forms – and used Ford or Perkins power plants. Cabcraft supplied cabs for the later models, which included air-conditioning.

Sadly the folding header idea was not taken any further and owners of other combine makes had to make use of removable headers that were taken off and towed behind the combine for transport, a system that is still universally used today.

Bizon, a Polish firm, had built its first combine in 1954, and during the 1970s sold in the UK for a time, including models such as the large Bizon Gigant with headers up to 26 feet wide, hydrostatic transmission and a 220 hp engine. Sales in Britain dwindled in the 1980s but the firm still supplied combines to Eastern Europe until it was taken over by Fiat in 1998.

Another European manufacturer that had a presence in Britain was Fortschritt from the former East Germany. This company built its first combine in 1948 and the E series models were offered in the UK during the 1970s, imported by Bonhill Engineering of Norfolk.

Massey Ferguson needed to keep ahead of the competition by providing larger and larger combine harvesters with bigger and bigger appetites to work on the huge, wide open fields of the Midwest in the USA. To this end the firm began work on two very large machines that would begin a whole new era of super combines. The 750 and 760, launched in America in 1971, also broke new ground in being supplied from new with a factory-fitted cab with air-conditioning. Both were very large machines, the 760 being the largest combine ever constructed so far. Its dimensions were certainly impressive, with a large diameter drum, a 24-foot cutting width and an output of up to 18 tons and hour.

Keeping with the original design of Massey-Harris combines since the 21, the 750 and 760 both had the driving position mounted to the left of the machine with the engine, a V6 unit supplied by Perkins, mounted beside it. Behind the engine and cab was the large grain tank, and as usual straw walkers were positioned behind this to move the straw to the very back of the machine.

Built at the large factory in Brantford, Canada, both new machines were very successful, especially with the custom cutters who soon built up large fleets of both models. They were also sold in Britain and Europe, but were not suited to the moister crops over here. Modifications, such as fitting the Multi-Flow re-thresher at the rear, helped improve their performance somewhat and the 765 model was designed for European conditions, despite still being built in Toronto along with the rest of the 700 range.

In Europe combines were still being produced in both France and Scotland, with the 520, 525 and 620 and 625 arriving in the 1970s. All were pretty large combines output-wise for this side of the Atlantic and the '5' once again denoted the presence of Multi-Flow. The 525 was probably the best seller in the UK.

What the 750 and the 760 did do was prove that there was a need for large super-size combine harvesters. Their sheer size put them in a class of their own and they are responsible more than any other machine for beginning the trend for ever larger and more efficient combines; a trend that continues to this day.

The Claas Senator replaced the Matador from 1966. The most obvious change was the new enclosed tinwork, while header widths were up to 20 feet wide. Power came from a 105 hp diesel engine while the cab was built by CabCraft. (Photograph: Kim Parks)

The Mercator and Protector versions of the Senator were later replaced by a range of Senator combines, including the Senator 70, which was available with either 105 hp six-cylinder Mercedes-Benz or Perkins diesel engines.

First seen in 1970, the Dominator would prove a hugely successful combine range for Claas. With new features and larger grain tanks, the Dominator was a popular choice for many. This 85 is harvesting peas in Suffolk.

The Dominator saw the introduction of a new comfortable cab with external lines that matched the square design of the combine itself.

The first of the new-style Dominator combines arrived from 1978 as the 6 Series. Replacing the original 66 model, the 68 was a later edition.

Biggest of the new Dominator range was the 106, powered by a 170 hp Mercedes-Benz engine and with a header width of 16.7 feet.

The new cab fitted to the 6 Series Dominators was an integral part of the combine's design, with built-in air-conditioning.

The controls in the cab of the Dominator 106 look basic but at the time were among the most advanced in the industry.

Oats pour from the unloading auger of a Dominator 106 as the 6,500-litre grain tank is emptied.

From the rear it is clear that the 106 was a large machine for its time. The straw chopper folded up neatly at the rear when not in use.

Although still in working position, it is obvious that the header and reel of the Lely Victory is made of two sections. Both fold up vertically for transport.

This Lely Victory was fitted with a CabCraft cab late in its life. Dating from 1977, it is seen harvesting winter barley.

Above: This is a later 1981 Lely Victory; header widths were from 14 to 18 feet while engine options included Perkins and Ford six-cylinder power units.

Left: Taken in the 1970s, this picture shows an International Harvester 953 combine at rest. Cabs were now fitted to these combines but they were never as popular in the UK as they were in other European countries.

The Super Cavalier was the very last model of combine to be built in Ipswich by Ransomes, with the last one being completed in 1976.

Laverda combines built in Italy were sold in the UK during the 1970s by Bamford, with this version being marketed as the Bamford Landlord.

East German combine maker Fortschritt sold several of its E Series combines into the UK via its importer Bonhill Engineering from 1989, including this E514 model. The first Fortschritt combine arrived in Britain as early as 1974.

The John Deere 900 Series combines were introduced in 1972 and were built in Zweibrücken in Germany. A later upgrade saw the original 900 Series updated into machines such as this 955.

Dating from the late 1970s, this John Deere 955 is harvesting oats in the south of England.

The Massey Ferguson 750 and 760 combines were launched in North America and were built in Brantford in Canada. From 1973 examples of both were sold in the UK, where this 750 is shown harvesting barley.

The 750 was a large combine for its day and the 760 was simply a giant. This example is shown in Suffolk, ready for another day's harvesting with its header on its trailer behind. (Photograph: Paul Reeve)

The Massey Ferguson 525 was a very popular combine in the UK during the late 1970s. It was powered by a 104 hp Perkins six-cylinder engine.

Most 525 combines were not fitted with a cab from new and so the operator often needed to wear protection from all the dust.

CHAPTER 5

Rotary versus Straw Walkers

From now on the size of combine harvesters would grow exponentially, both in terms of physical size and in cutting widths. Straw walkers were still king, being fitted after a conventional drum and concave, and firms such as Massey Ferguson kept with this principle for some time, notably with the Brantford-built 850 and 860 models that followed the 750 and 760 machines, which were sold in Europe as the modified 855 and 865 combines. Sadly, combine production in Kilmarnock came to an end in the early 1970s, but a new 800 Series range was still produced in France based on the earlier models, and included enough improvements to see the range into the 1980s.

Other manufacturers looked at ways of improving the very concept of how the combine harvester worked. International Harvester spent a great deal of time and money developing the Axial-Flow principle, a machine that did not use a conventional threshing system and straw walkers at all. Instead, it was fitted with a large single rotor that did the threshing and was also extremely efficient in the growing number of different crops that combines were now being asked to harvest, such as rice and maize.

In 1977 International Harvester launched a range of Axial-Flow combines, which were an instant hit. They looked different as well, with their sleek styling and a shorter overall length thanks to the lack of straw walkers. They were not to everybody's taste, but they certainly worked well enough in most conditions, especially in the USA. In Europe they also sold well and eventually a factory was set up in France to build them. In Europe, however, straw was seen as a much more important commodity than in the USA, were it was usually chopped behind the combine. Many farmers were not happy with the way the rotary Axial-Flow damaged the straw, making it more difficult to bale and store for cattle bedding.

The original Axial-Flow 1400 Series was made up of three models and these were gradually improved over the years, evolving into the 1600 Series in 1985, by which time International had been merged with Case to form Case IH.

New Holland was also busy working on a rotary principle for combine harvesters and came up with the Twin Flow rotary system, which was first seen on the American TR range in the late 1970s. This technology was evolved to also be included in the European

TF range by the 1980s and enjoyed a great deal of success. The European TX range was the straw walker equivalent and was very popular in much of Europe, including the UK. These replaced the early 8000 Series, which had been the first to be fitted with the firm's own factory-fitted cabs. Built using bonded glass on the front and sides, they provided excellent visibility of the work at hand. A similar but larger cab was fitted to the TX and TF combines before a revamp saw the Discovery cab fitted, which featured a curved front screen and many electronic functions to control the machine.

The Ford Motor Company bought New Holland in 1985 and the company became known as Ford New Holland, and remained so for a time after Fiat took over from 1991.

Massey Ferguson, steadfastly dedicated to the straw walker principle, eventually relented and introduced a rotary combine, but this was done by acquiring the White combine harvester business and its new rotary model. The result was the huge 8590, a massive machine with a huge appetite better suited to North American conditions than European. Similar to the Axial-Flow in many respects, these rotaries used a single large rotor, and although trials in Europe were carried out they remained largely a North American phenomenon.

Massey Ferguson had other ideas up its sleeve, though, when it acquired the Dronningborg factory in Denmark. This firm had been making pretty advanced combine harvesters for many years for the European market, and had recently been doing a lot of work with electronic control and monitoring systems.

It was this that attracted Massey Ferguson to take a large stake in the Dronningborg business in 1984, based in Randers, and soon a range of straw walker combine harvesters were being produced in MF colours for sale around the world, but particularly in Europe. An electronic monitoring system was a great advance in combine harvester operation and this was gradually improved over the years to include a full computer-controlled system, which by the 1990s could be linked to satellites to provide yield mapping of individual fields.

Another Massey Ferguson innovation of this time was the Powerflow header, which used a conventional style header incorporating a belt to allow the crop to be presented to the intake auger in a uniform fashion and improve efficiency. All the combines in Denmark remained straw walker machines, but a rotary separator was added to give them better threshing capacity.

Deutz can trace its history back to 1894, while Fahr was founded in the 1870s. Both German firms joined forces in 1961, the former specialising in air-cooled engines and tractors while the latter was a farm implement producer, although Deutz had made threshing machines early in its existence. The first Fahr combine appeared in 1951, with development continuing apace after the formation of Deutz-Fahr. Sticking to conventional straw walker type machines, Deutz-Fahr enjoyed much success in the UK, first with the M Series and then the Topliner range. An attempt to market a massive eight-straw walker giant was not successful and combine manufacture ceased for a time after the Italian-based Same Lamborghini Hurlimann group acquired the business.

Claas, meanwhile, had been developing its combine range over the years, gradually offering larger and larger models and also a new cab when the 8 Series arrived from 1981. A version of the Dominator design was produced called the Commander, allowing Claas to enter the rotary market, but the straw walker Dominator range was certainly the best seller of the two, the rotary concept being slower to take off in Europe than in the USA.

The Lexion 480 combine of 1996 changed all that though. This brand-new design, with the new stylish Vista cab, featuring curved glass and a computer monitoring system, was a revolution in combine design and was a rotary model with a huge appetite. Gradually all the Dominator range was replaced by the Lexion models, in both rotary and conventional form.

John Deere was building combines in both North America and in Germany, although often to different designs. In the USA large Titan-badged machines were built to take on the might of the MF 760, but in Europe the John Deere designs remained fairly conservative. Along with others of its range, the 1075 broke new ground with the new John Deere SG cab. Hydrostatically controlled combines were also growing in popularity with users, and John Deere produced a Hydro range to keep up with the competition. Hydrostatic drive made operation much simpler as it replaced a manual gearbox and allowed forward and reverse to be controlled by simply moving a single lever forwards or backwards to increase or decrease speed.

When the Z Series arrived in 1993 John Deere combines were among the largest and most powerful around, with sleek styling and advanced features, including their well-appointed cabs. Versions of these were built in both Germany and in the USA, although those in America were always fitted with wider headers than in most of Europe. The largest was the 2066 model, which came with a high level of automatic adjustments. In 1997 the improved 2200 Series replaced the original range, with the 2266 now being the largest.

A rotary model finally arrived from John Deere in the shape of the CTS, a combine originally designed for harvesting rice but adapted to become a rotary jack-of-all-trades machine. CTS stood for 'Cylinder Tine Separation', which was the type of threshing system used. It was successful to a point, but it took further development to become a truly versatile rotary machine.

The battle between straw walker and rotary combines continues to this day, with all the major manufacturers offering a choice of type, but perhaps the main growth has been in the area of the hybrid – a combine that uses the best of both systems in one machine. This means that nowadays the farmer and contractor have no less than three main systems to choose from and can tailor their choice to the conditions and crops in which they expect their new machine to work. In many ways all the various types of threshing system have evolved to such a high standard that any of them can be used as a universal machine capable of dealing with around 100 different crops, and the choice of which type to use is simply down to the purchaser's preference. Even the original problem of rotary models requiring more power to operate is less of a problem today as large combines with very powerful engines are the norm, meaning that any power loss experienced on a rotary machine is now fairly negligible.

A Massey Ferguson 850 harvesting maize in the Midwest. Built in Canada, the 850 replaced the 750 in 1981. (Photograph: Paul Reeve)

The biggest combine available from Massey Ferguson in the early 1980s, the 860 was powered by a 184 hp Perkins V8 engine. (Photograph: Paul Reeve)

The 855 was primarily designed for European conditions and incorporated a cascade separator system to avoid grain losses. (Photograph: Kim Parks)

Inside the cab of a Massey Ferguson 865 it can be seen how much more refined the actual controls now were compared with earlier machines.

The 825 was built in France from 1983 and was based on the earlier designs with several improvements, including a new cab. (Photograph: Kim Parks)

International Harvester changed everything with the development of the Axial-Flow rotary combine. First launched in 1978, this is a second-generation machine – a 1640 model introduced in 1986.

Two generations of New Holland TR combines in North America. The Twin Rotor principle, introduced in 1975, proved not to be ideal for European conditions. (Photograph: Paul Reeve)

The New Holland 8080 was the largest in the 8000 Series launched in 1976, and was the first five-straw walker machine built at the Belgian factory.

The mid-1980s saw the arrival of the New Holland TX range of large straw walker combines, with the TX36 being powered by a six-cylinder Ford engine.

New Holland launched the TF42 and TF44 in 1983, bringing the rotary concept to European fields with the Twin Flow concept. Crops such as oilseed rape could be harvested at a faster rate than with a straw walker machine.

A new Discovery cab was a feature of the New Holland TX65.

The ultimate evolution of the TF rotary range was the TF78 Elektra, which was first seen in 1995. This large machine was extremely advanced, with fibre optic cables being used for the electronic control and monitoring systems.

Emptying on the move shows the large size of the flagship New Holland TF78 Elektra.

By the 1990s, all Massey Ferguson combines in Europe were built in the former Dronningborg factory in Denmark, including the mid-sized 34.

The Massey Ferguson 38RS included a rotary separator as well as straw walkers. This demonstrator is shown detaching its header before moving on to the next farm in 1991.

Deutz-Fahr introduced the M36.10 in 1985. It was available with mechanical or hydrostatic drive.

The Deutz-Fahr M36.40 was a later addition to the range and was powered by a six-cylinder Deutz engine.

The Topliner range was launched in 1991 and included the 4075HTS model, with six straw walkers and a 240 hp six-cylinder Deutz engine.

The Claas Dominator 98 arrived in 1985, powered by a 150 hp Mercedes-Benz engine.

The 98S version of the Dominator had manual drive but was otherwise the same as the standard model.

A close-up of the new Claas cab fitted to the 8 Series Dominators with its large glass area and improved driver comfort.

The Claas Commander 115C was launched in 1986 and used a new eight-cylinder rotary design, making it shorter than the Dominator range as there were no straw walkers. Power came from a 250 hp Mercedes-Benz engine, the threshing system requiring extra grunt.

The Lexion 480 was the very first of the hugely popular Lexion range from Claas and first appeared in 1996. This was a twin rotor machine with a mighty 375 hp engine, a large 10,500 litre capacity grain tank and a new Vista cab.

The 480 was large with a 7.5-meter-wide header and was the first 'hybrid' combine, using a conventional threshing drum and rotors.

In Europe the 1100 Series John Deere combines were very successful from 1987. The improved 1100 SII versions arrived in 1990, with the 1175 SII being the mid-sized model. (Photograph: Kim Parks)

First seen in 1993, the John Deere Z Series was topped off by the 2066. These big combines were powered by six-cylinder Deere power plants.

This John Deere 2066 is harvesting swathed oilseed rape with a special header attachment. (Photograph: Paul Reeve)

CHAPTER 6

Big Is Beautiful

All the major combine harvester manufacturers have built machines of all sizes, to appeal to farms of all sizes, but the ever-increasing trend has been for even larger and more efficient machines.

Efficiency has been made possible not just by the large size and wider headers, but also by the amount of electronic control and fine-tuning available in the modern machines. Computers continually monitor all the operations of the combine, including grain loss, and constantly adjust the machines settings automatically. Self-driving systems have been developed using laser-guided cameras, and more recently GPS systems that use satellites to position and guide the combine across the field in perfectly straight lines, without input from the driver needed after initially setting the machine in motion.

The Claas family of Lexion combines has evolved enormously since the name was first used. These are the largest combines built by this German manufacture, which now also builds combines in the USA for the North American market as well as Germany. This is a remarkable instance of a European manufacture turning the tables and sending its combine harvester technology from Europe to the USA – the complete opposite of the direction of evolution previously seen in the story of the combine harvester.

The Lexion 600 was one of the world's largest combines when launched in 2006. It came with a 30-foot-wide header and was bristling with electronics, including auto-guidance systems and fully automatic monitoring aids. Soon 35-foot-wide headers were made available, making more use of the power available from the V8 Mercedes-Benz 586 hp engine, and 40-foot-wide headers were to follow, such was this machine's huge appetite for cereal crops. The 600 was also often specified with the Terra-Trac rubber track system option, which replaces the front wheels. Tracks are not a new idea for combines, as right back in the early days machines such as the Massey Harris 21 were equipped with metal tracks in place of the front wheels to help them cope in flooded paddy fields when being used to harvest rice. The downside with metal tracks was that they damaged road surfaces when travelling from field to field, and so they needed to be transported by truck. However, Caterpillar developed a rubber track system for crawler tractors at the end of the 1980s, and this technology has now been adapted to make a more versatile form of go-anywhere combine harvester by all the major manufacturers.

The Claas Lexion 780 is the largest production green machine so far, taking the success of the 600 and improving on that formula, with 45-foot-wide headers often being the norm.

New Holland, part of Ford New Holland from 1985, was taken over by Italian firm Fiat in 1991 and merged with the Laverda combine business to produce even more advanced machines, especially in the area of hillside combines. The idea of a hillside combine is that hydraulically controlled rams allow the main body of the combine, including the threshing drum, to remain level even if the wheels and header are following steeply sloping land. This means that threshing efficiency remains at its best as the drum keeps level. If tilted then grain can be lost, but this ingenious system, actually used on some early North American reaper-threshers but in a less sophisticated form, prevents this from happening.

Since the takeover by Fiat, whole new ranges of combine harvester have been produced by New Holland, including rotary and conventional, and incorporating all the modern features of auto-guidance, computer control and monitoring, as well as some futuristic-looking cabs. Two main ranges were produced, with the CX being based on the older TX threshing principle using straw walkers, while the CR models were rotary combines. The largest today is the mighty CR10.90, which is a true 'super combine' in the mould of the Massey Ferguson 760 of the 1970s – but much bigger!

Case IH also became a part of Fiat in 1999, but has largely kept its identity separate, especially when it comes to its Axial-Flow rotary combine range, which is still going strong. In fact, the Case IH brand was almost exclusively rotary for many years such was the success not only of the rotary concept, but the Axial-Flow brand itself.

John Deere have embraced rotary technology as well as conventional and hybrid designs to produce a wide range of machines. These offer a high level of technology, with different features to suit all types of harvesting conditions and crops on a global scale, and the green and yellow combines are a familiar sight in harvest fields everywhere – but perhaps no more so than in the USA, where John Deere holds the dominant position once held by Massey Ferguson for use by contractors on the great plains.

Massey Ferguson is now part of the large AGCO conglomerate, which also includes Fendt, Gleaner, Valtra and Challenger, among others. This has led to a very complicated evolution of combine designs, especially in the USA. Here in Europe this has been marked by the Dronningborg-designed Massey Ferguson machines being badged as Fendt and painted in this firm's green colour scheme as well as the familiar MF red. Eventually the factory in Denmark would close, but both red and green combines are still produced.

In the USA the situation is more complex. For many years the Gleaner combines were extremely popular and stood out from everything else as they were finished in galvanised steel and not painted. The Gleaner marque is an important one and the make has always been popular in the Midwest in particular. The firm was once part of Allis-Chalmers and is today part of AGCO.

The Challenger brand is also part of AGCO and started with the acquisition of the Challenger crawler tractor range from Caterpillar. Caterpillar had also entered into a marketing arrangement with Claas to provide CAT-badged Lexion combine harvesters in the States, but this was not part of the AGCO deal, leading Claas to set up its own distributer network in North America. AGCO then used Massey Ferguson-based designs and branded them as Challenger, which had been developed from earlier North American machines rather than based on European combines.

So with all these different brands, AGCO has a large share of the combine market, especially in the USA. Recently the firm has announced the massive Delta combine range, which will be only available in black livery with Massey Ferguson, Fendt or Challenger branding.

Using the Claas Roto-Plus hybrid threshing system, the Lexion 580 Terra-Trac is a large combine with a 462 hp V8 engine, a grain tank capacity of 10,500 litres and header widths up to 30 feet.

An unusual sight in the UK, this Claas Lexion 580 Terra-Trac is harvesting maize with a special header.

Appearing in 2005, the Claas Lexion 600 was a true super combine, with headers up to 35 feet in width, a massive 12,000-litre grain tank and a 586 hp Mercedes-Benz V8 engine.

Inside the Claas Vista cab of the Lexion 600, the operator is comfortably seated with ergonomically located controls.

From 1998 Lexion combines were sold in the USA under the Caterpillar brand. This one is fitted with a Shelbourne Reynolds stripper header that only harvests the grain heads. (Photograph: Paul Reeve)

The Claas Vista II cab fitted to this Lexion 770 – part of the latest upgraded Lexion range – provides even more operator comfort than before.

Claas claimed the massive Lexion 780 had the biggest output of any combine in the world.

New Holland replaced the TF range with the CR rotary machines in 2002. Radical new styling made them stand out, with the CR980 being the largest, power coming from a 428 hp Iveco engine.

The New Holland CR range was built both in Belgium and in the USA and evolved into the CR9000 Series, including the CR9070.

The CX range replaced the New Holland TX line-up from 2002 and included the new Discovery II cab.

In 2006 New Holland launched the CSX Series. The CSX7080 was the largest with a 333 hp engine and a Discovery Plus cab.

New Case IH Axial-Flow models arrived in 1995 with a new cab and more electronic controls. The 2166 was the second largest in the original line-up.

The Case IH 2300 Series Axial-Flow range arrived in 2002 and this 2366 is seen harvesting maize game cover.

The 2377 Axial-Flow was originally designed for just the French market, but this one made it to Britain and is shown at work in Suffolk.

The Case IH Axial-Flow 5088 was the smallest in a new range launched in 2008, and came complete with more curved side panels designed for easy access.

The 20 Series Axial-Flow combines were launched in 2007 and took the principle to new heights. They later evolved into machines such as this massive 9230.

Clearly evolved from earlier models, the John Deere 9580WTS was a mid-range model using the WTS system, an improved straw walker threshing method.

John Deere rotary combines were still fitted with the CTS system.

The John Deere cab has evolved to keep up with the requirements for operator comfort as well as to provide an unobstructed view of the header.

The John Deere 9580iWTS is fitted with a completely integrated electronic system of self-steering and other automatic features.

The largest machine in the John Deere line-up was the S690i, with the Single Tine Separation system using a single large rotor for threshing. Rubber tracks are now also an option. (Photograph: Kim Parks)

The largest of the European range for many years, the Massey Ferguson 40 was powered by a 300 hp Valmet engine and was replaced in 2000 by the new 7200 range, which was still built in Denmark.

In the USA, Massey Ferguson combines were purely rotary and large, as seen by this machine on display at a trade show. As part of AGCO, components are now shared between the brands Gleaner, Challenger, MF and Fendt. (Photograph: Paul Reeve)

CHAPTER 7

Future Combines

Combine harvesters have come a long way in the hundred years or so that they have existed, from early machines made out of a lot of wood and pulled by horses, to the massive leviathans filled to the brim with electronics and computer-controlled functions that we see today.

As to what the future holds, well they will no doubt continue to get bigger, although the 45-foot-wide headers around today are probably getting towards the limit of what a conventional combine with its comparatively narrow feeding auger can cope with, at least in thick European crops. Other designs have been mooted and some are in production, such as the intriguing Tribine, which is a large articulated monster.

Perhaps it will be a case of smaller machines being used as a swarm. The latest drone technology allows one operator to control two or more vehicles, and this could be applied very easily to a combine harvester, the extra cutting width being gained over three machines or more, all controlled by just one person. Of course autonomous vehicles that don't need the input from an operator at all are also being developed, so perhaps in the future humans will not be needed and robot combines will be going around by themselves, harvesting the crops as soon as they sense they are ripe.

It seems inevitable that computers will take over more as time goes on and automation will certainly be the thing of the future, but as to the actual workings of the combine, the chances are that it will still work in much the same way as previously, probably by more refined hybrid-type technology and materials, but basically still threshing out the grain from the straw as combine harvesters have always done.

One thing is certainly for definite – the world will always need and rely on the combine harvester; probably the most important machine ever built by mankind!

A Claas Lexion
780 Terra-Trac gobbles
up a wheat crop.

Special headers
are used for maize
harvesting in the USA
and in Europe, as
shown on this Case IH
Axial-Flow.

Very wide headers are
now the norm, as are
rubber track units.

John Deere uses taller tracks compared to those used by Claas, but both systems provide greater ground contact and less compaction.

A New Holland TX66 harvesting wheat into the sunset.